Shop Safety Workbook

Student _____

School _____

ISBN 978-0-6151-4892-2

Schirmer Computer Services
1018 Aaron Rd.
Summerville, PA 15864

Notice to the User

The publisher and author does not guarantee or warrant any of the procedures or products contained herein. The user is expressly warned to consider all products and procedures as extremely dangerous and extreme caution must be exercised. This workbook cannot and does not warranty or guarantee any process contained herein. The user willingly assumes all risks in connections with these instructions. The publisher shall not be held liable for any special consequential, or exemplary damages resulting, in whole or in part from the users use of, or reliance upon this workbook.

Copyright 2006

All rights reserved. No part of this work may be reproduced or used in any form by any means including graphic, electronic, mechanical, including photocopying, recording, taping, internet posting, or information storage or retrieval systems. This workbook is protected under the Copyright Law of the United States.

ISBN 978-0-6151-4892-2

All photographs and layout by Grant W. Schirmer

Introduction and how to use this workbook:

As a vocational agriculture teacher for 25+ years, I have noticed a need for a well thought out safety workbook for students to use. As department chairman for technology education (formerly industrial arts) and vocational agriculture in my school district, my shop teachers have often said "Why can't there be a workbook for every student, and each year they go through the lessons, and we would have safety records for each student. At the end of the year the student could take the workbook home as a working safety text for reference." I have noticed accidents in schools that possibly could have been prevented with more safety instruction. In today's litigious society, there can never be enough safety instruction and anecdotal record keeping.

This workbook is the result of many years of refining safety education. The machines in this book are typical school shop machines. They are real machines, in use everyday, by real students, in a real school setting. They are the same types of machines used everyday in industry.

To use his workbook effectively, the teacher should issue a workbook to each student. The teacher explains the machine, then explains each major machine part in this workbook. The teacher then has the students read aloud each major safety rule and discuss each of them. The teacher then takes the students to the shop and demonstrates the machine and it's safe operation. The teacher then observes each student doing a simple task on each machine and checks that proficiency off on the power tool exercise sheet. The students then may view any additional machine information i.e. school textbooks, internet research, videos, etc. The students then complete the review questions and safety quiz and sign the machine understanding sheet. Upon completion of study of all the machines in the school's shop, the teacher reviews all the machines and gives a final safety examination.
This workbook is kept on record and at the end of the year, issued back to the student and they may take it home for reference.

I have used this system for many years with great results. Sometimes I have students for 4 years and they go through this drill down each year. Safety cannot be overemphasized and not over taught. I stress to my students, when there is a safety situation, your training takes over. It is a skill learned through constant repetition, teacher observation and guidance. This work book is a great starting point.

David Schirmer
Vo Ag Teacher
Author and Publisher

Contents

1. General Lesson Plan p.6
2. Shop Organization p.7
3. Fire Safety p.8
4. Metal Lathe p.9
5. Drill Press p.11
6. Table Saw p. 13
7. Jointer p. 15
8. Radial Arm Saw p.17
9. Band Saw p. 19
10. Scroll saw or Jig Saw p. 21
11. Grinder p.23
12. Power Hack Saw p. 25
13. Power Miter Box p.27
14. Portable Hand Drill p. 29
15. Router p. 31
16. Portable Belt Sander p. 33
17. Portable Electric Saw p. 35
18. Planer-Surfacer p. 37
19. Arc Welder p. 39
20. Oxyacetylene Welding Equipment p. 41
21. Disk Sander p. 43
22. Vertical Milling Machine p. 45
23. Horizontal Band Saw p. 47
24. Common Hand Tools p. 49

Machine User Demonstrated Safety Check Sheet

Machine	Student Usage (teacher initials)	
Metal Lathe		
Drill Press		
Table Saw		
Jointer		
Radial Arm Saw		
Band Saw		
Scroll Saw or Jig Saw		
Grinder		
Power Hack Saw		
Power Miter Box		
Portable Hand Drill		
Router		
Portable Belt Sander		
Portable Electric Saw		
Planer-Surfacer		
Arc Welder		
Oxyacetylene Welding Equip.		
Disk Sander		
Vertical Milling Machine		
Horizontal Band Saw		

Teacher Sample Lesson Plan

Unit: Shop Safety
Lesson: Drill Press Safety
References:
This workbook and any other relevant materials.
Setup: Prepare the machine and have scrap wood for the students to practice to drill through.

Teaching Procedure:

The teacher should issues a workbook to each student. The teacher explains the machine, then explains each major machine part in this workbook. The teacher then has the students read aloud each major safety rule and discuss each of them. The teacher then takes the students to the shop and demonstrates the machine and it's safe operation. The teacher then observes each student doing a simple task on each machine and checks that proficiency off on the power tool exercise sheet. The students then may view any additional machine information i.e. school textbooks, internet research, videos, etc. The students then complete the review questions and safety quiz and sign the machine understanding sheet. Upon completion of study of all the machines in the school's shop, the teacher reviews all the machines and gives a final safety examination.

This workbook is kept on record and at the end of the year, issued back to the student and they may take it home for reference.

Evaluation:

The student is observed and checked on the machine's correct usage. The student completes a quiz and a final test.

This workbook addresses:
Pennsylvania Standards for Science and Technology 32.7B, 37.7A, 37.10A, 37.12A, 36.7C, 36.10C, 36.12C

Shop Organization

Shop organization is a very important aspect of the safety training you will receive. All tools, machines and supplies must be cared for properly. General shop safety is the responsibility of everyone. A very efficient system to use is the shop foreman system. It is patterned after industry models. The class chooses individuals for these tasks or the teacher assigns them these responsibilities.

TITLE	SUGGESTED DUTIES
General Foreman	Supervise shop clean up, supervise overall shop organization and safety, act as a student representative to the teacher. Report to the teacher at the end of class.
Stationary Power Tool Foreman	Check power machines for safety and operation, check blades for sharpness, bring any maintenance issues to the teachers attention, report to the General Foreman at the end of class
Portable Power Tool and Hand Tool Foreman	Check hand tools and hand power tools are put away and in good safe working order, bring any maintenance issues to the teachers attention, report to the General Foreman at the end of class
Project Foreman	Make sure all projects identified are stored properly, bring any maintenance issues to the teachers attention, report to the General Foreman at the end of class
Safety Foreman	Corrects and reports any unsafe condition to the teacher, check safety goggles-condition and count, safety signs and posters, bring any safety issues to the teachers attention report to the General Foreman at the end of class

The last 10 minutes of the lab period is used for clean up and safety inspections daily. All Foreman report to the General Foreman who reports to the teacher before class is dismissed.

Fire Safety

All students should follow fire and emergency drills as prescribed by your school. Exits should clearly marked. The shop should be free of combustible materials. To understand fire safety, students should understand the fire triangle. Fuel, Heat, and Oxygen are the components . If you remove any of these components a fire cannot be started and with the removal of any one the fire will be extinguished.

Fire Classes

Class A– Ordinary combustibles including paper, wood and trash

Class B-Flammable liquids including fuels, grease, paints, etc.

Class C-Electrical equipment

Class D-Combustible metals

The teacher should show the class where the fire extinguishers are located, the types of extinguisher and demonstrate their usage. The class should know also the fire evacuation procedures in your school.

Metal Lathe

Safety Rules:

1. Operate only after you have received proper safety instruction from your teacher.
2. No loose clothing.
3. Always wear goggles.
4. Remove the tool holder for polishing.
5. Remove the chuck key prior to operation.
6. Remove all chips with a bench brush.
7. Make sure all pieces are secured.
8. Be sure floor area around machine is clean and not slippery.
9. See your instructor for the maximum permissible cut allowed on your machine.

Lathe

Review Questions:

1. Why is it important to not have loose clothing around a lathe?
2. What is the maximum permissible cut on your lathe?
3. Why are goggles necessary?
4. Why should the cutting tool be kept sharp?
5. Describe how to chuck a piece in the lathe chuck.
6. How do you face off a piece?
7. Who manufactures your machine?

Safety Quiz: True or False

1. Goggles are not necessary when using this machine.
2. It is a good idea to jam the cutting bit into the stock.
3. The floor area should be clean around the machine.
4. A loose T shirt is OK when operating this machine.
5. Running a lathe requires your undivided attention.

Teacher Assignment:

I understand the lathe.   Signed:_____
Date_____

Drill Press

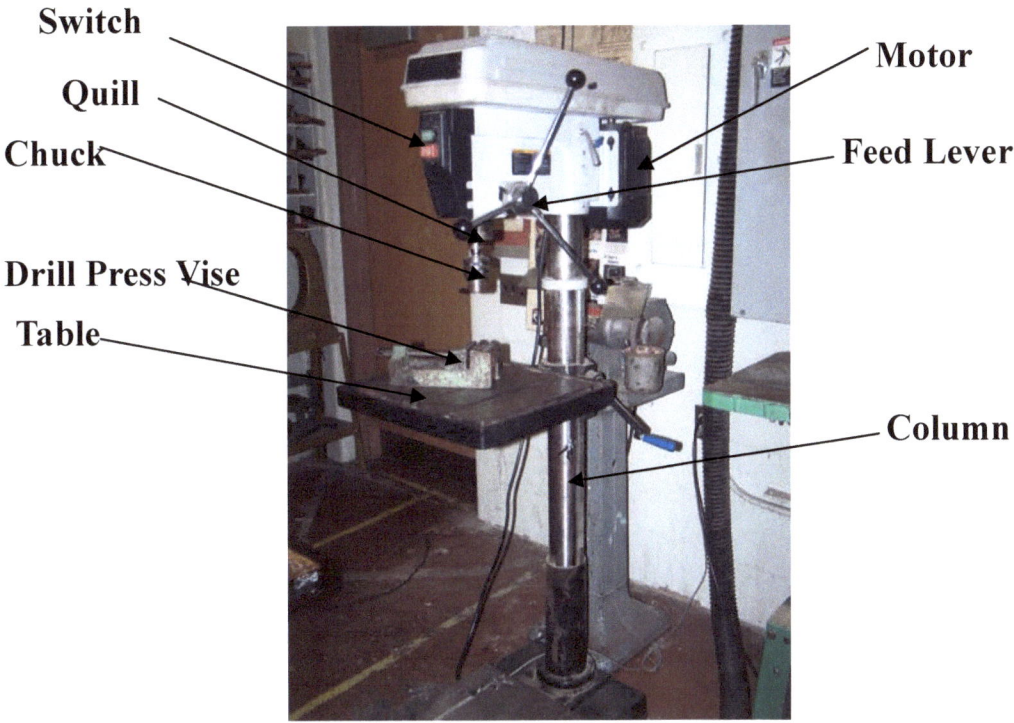

Safety Rules:

1. Operate only after you have received proper safety instruction from your teacher.
2. No loose clothing.
3. Always wear goggles.
4. Pieces being drilled must be clamped or held by a vise.
5. Remove the chuck key from chuck before starting the machine.
6. Remove all drill chips with a bench brush.
7. Never talk to a machine operator. Running a machine requires your undivided attention.
8. Be sure floor area around machine is clean and not slippery.
9. Never leave an unattended machine with the power running.
10. Be sure the twist drill does not drill into the drill press table.
11. Use only sharp twist drills.
12. When drilling, let up on the feed lever just prior to the twist drill "breaking through". This can prevent the twist drill from "catching" or the bit seizing.
13. Use the proper speed for the material being drilled. Softer materials can be drilled at higher speeds, while harder materials require a slower speed. Check the manufacturer's recommendations. A drilling lubricant may be required.

Drill Press

Review Questions:

1. Why is it important to secure stock when drilling?
2. Explain how to drill through a piece of stock.
3. Why should the twist drills be kept sharp?
4. Why is drill speed important?
5. Who manufactures your machine?

Safety Quiz: True or False

1. Loose, long hair around the chuck is safe.
2. Goggles are mandatory.
3. You can use the same drilling speed for all materials.
4. Oil around the floor is a good idea.
5. Holding pieces in your hand while drilling is OK.

Teacher Assignment:

I understand the drill press.   Signed:_____
Date:_____

Table Saw

Guard **Fence**

Table **Blade**

Switch

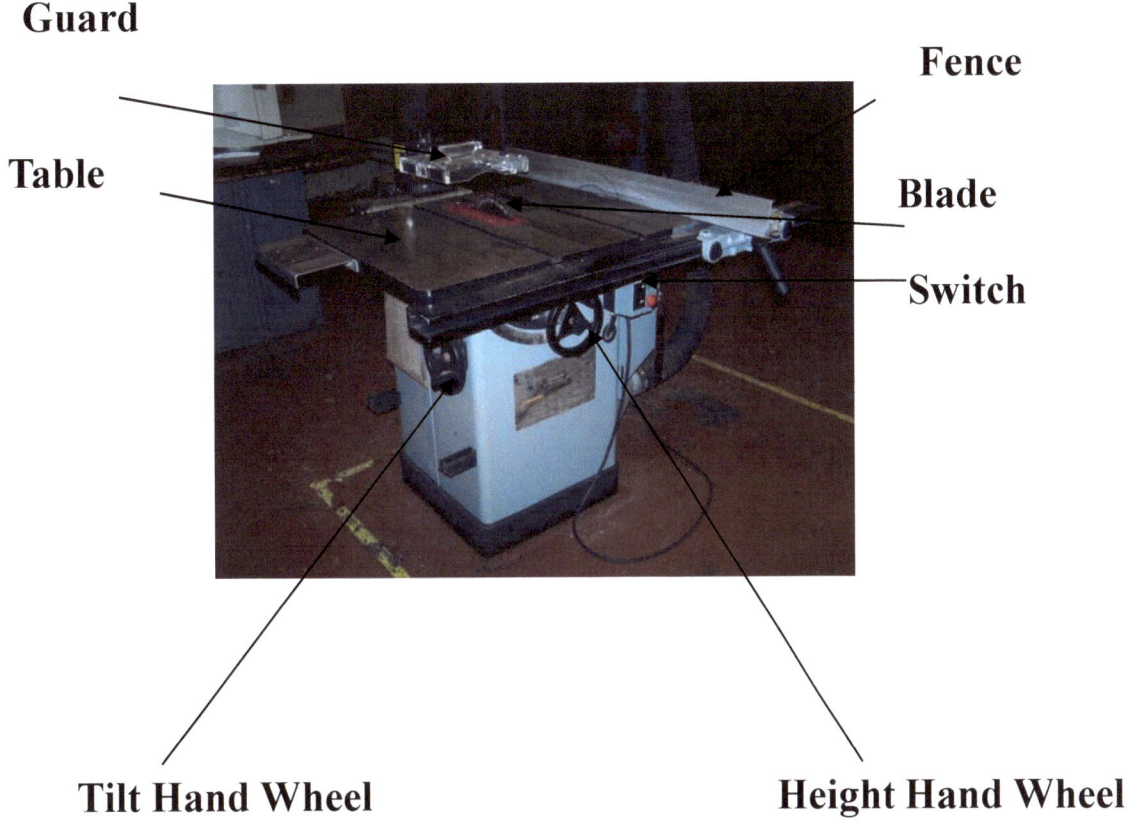

Tilt Hand Wheel **Height Hand Wheel**

Safety Rules:

1. Operate only after you have received proper safety instruction from your teacher.
2. No loose clothing.
3. Always wear goggles.
4. Always use the guard.
5. Blade height should not be more than 1/4" above the board.
6. Remove all scraps with a bench brush after machine is stopped.
7. Narrow pieces require a push stick.
8. Be sure floor area around machine is clean and not slippery.
9. Fence must be used for ripping– Miter Gauge for crosscutting
10. Fingers must always be clear of the blade.
11. Do not cut round stock on a circular saw.
12. All special setups must be checked by the instructor.

Table Saw

Review Questions:

1. Why is it important to have correct blade height?
2. Why must the fence be used when ripping?
3. Why must the miter gauge be used when crosscutting?
4. What is a push stick?
5. What machine could be used for cutting round stock?
6. Why is the guard important?
7. Who manufactures your machine?
8. What is kickback? How can it be prevented?

Safety Quiz: True or False

1. Goggles are not necessary when using this machine.
2. It is safe to crosscut with the miter gauge.
3. The floor area should be clean around the machine.
4. The fence is not necessary when ripping stock.
5. When ripping, it is safe to not push the stock clear of the blade and just let it go free.

Teacher Assignment:

I understand the table saw.   Signed:_____
Date:_____

Jointer

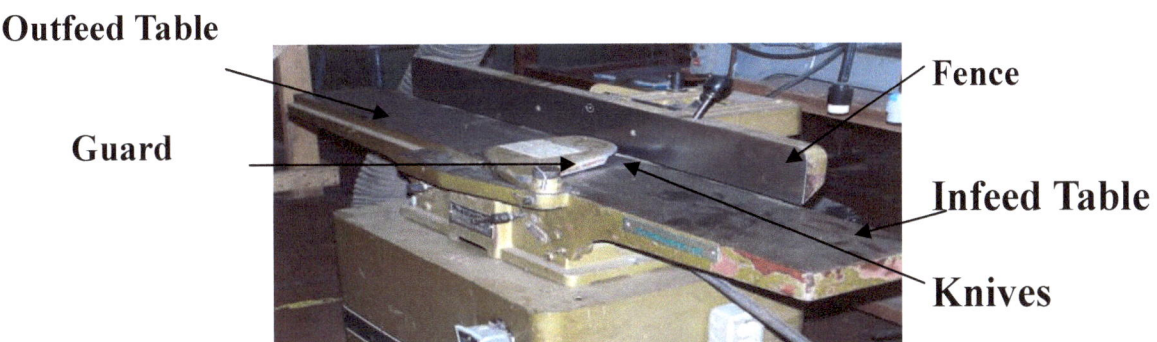

Safety Rules:

1. Operate only after you have received proper safety instruction from your teacher.
2. No loose clothing.
3. Always wear goggles.
4. The guard must be in place and working.
5. Use a push shoe for small pieces.
6. Remove all scraps with a bench brush
7. Special setups must be checked by the instructor.
8. Be sure floor area around machine is clean and not slippery.
9. See your instructor for the maximum permissible cut allowed on your machine.

Jointer

Review Questions:

1. Why do jointer knives rotate towards you?
2. What is the maximum permissible cut on your jointer?
3. Why does the infeed table have to be lower than the outfeed table?
4. Why should the cutter head knives be kept sharp?
5. Describe how to join the edge of a board.
6. Who manufactures your machine?

Safety Quiz: True or False

1. A jointer is used to drill holes.
2. Use a jointer as fast as possible.
3. Jointers can be used to joint edges of wood.
4. A jointer's guard must be working.

Teacher Assignment:

I understand the jointer   Signed:_____
Date:_____

Radial Arm Saw

Safety Rules:

1. Operate only after you have received proper safety instruction from your teacher.
2. No loose clothing.
3. Always wear goggles.
4. All adjustments are made with the power off.
5. All guards must be in place and functional.
6. Remove all scraps with a bench brush
7. Stock must be held securely against fence when crosscutting.
8. Be sure floor area around machine is clean and not slippery.
9. Ripping requires extreme care and should only be done under direct supervision of the instructor.
10. Always return the cutter head to the rear of the machine after the cut and secure if possible.
11. Go slow enough when crosscutting so that the blade does not "walk" up on the board and jam the machine.

Radial Arm Saw

Review Questions:

1. Why must the cutter head be returned after each cut?
2. Why must the guard work properly?
3. Explain why extreme care must be exercised when ripping.
4. Describe how to crosscut a board.
5. Who manufactures your machine?

Safety Quiz: True or False

1. It's OK if the guard sticks on this machine.
2. The fence is very important.
3. Crosscutting can be done on this machine.
4. Make adjustments on this machine while it is running.
5. This machine does not have to be at full speed when using it.

Teacher Assignment:

I understand the radial arm saw. Signed:_____
Date:_____

Band Saw

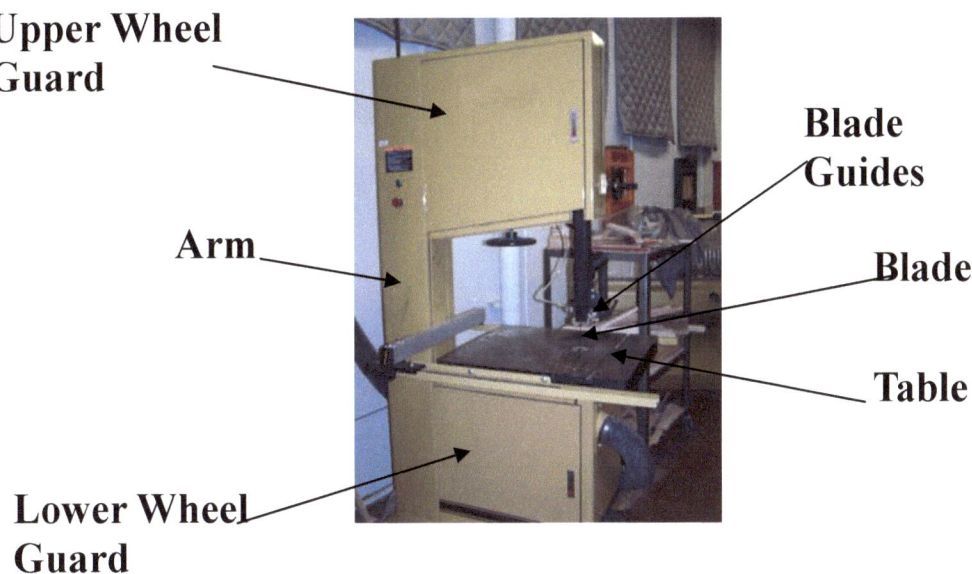

Upper Wheel Guard
Arm
Lower Wheel Guard
Blade Guides
Blade
Table

Safety Rules:

1. Operate only after you have received proper safety instruction from your teacher.
2. No loose clothing.
3. Always wear goggles.
4. The top guide should be within 1/4" of the work.
5. The guide wheels should be barely turning.
6. Avoid backing out of a cut if possible.
7. Never twist the blade. It will overheat and may crack.
8. Be sure floor area around machine is clean and not slippery.
9. If the band saw is making a "knocking " sound the blade may be partially broken.
10. Use a "V" block for cutting cylindrical stock.
11. Keep fingers out of blade path.

Band Saw

Review Questions:

1. Why is it important to keep your fingers away from the blade guides?
2. Why is it a bad idea to back out of a cut?
3. Why can't you twist wood sharply when using this machine?
4. What does a knocking sound from the blade indicate?
5. Who manufactures your machine?

Safety Quiz: True or False

1. A band saw with the appropriate blade can be used to cut steel.
2. The top guide should be 1" from the piece being cut.
3. Goggles are necessary when using this machine.
4. It is a good idea to talk to others when using this machine.
5. If a blade breaks, alert your teacher immediately.

Teacher Assignment:

I understand the band saw. Signed:_____
Date:_____

Scroll Saw or Jig Saw

Safety Rules:

1. Operate only after you have received proper safety instruction from your teacher.
2. No loose clothing.
3. Always wear goggles.
4. Fingers should be always clear of the blade path.
5. Work should be held securely on the table.
6. Remove all scraps with a bench brush when the machine is stopped.
7. Blade tension should have a "musical" quality.
8. Be sure floor area around machine is clean and not slippery.
9. Teeth always point down on a jig saw blade.

Scroll Saw or Jig Saw

Review Questions:

1. Which way do the teeth point on the blade?
2. Describe your blade attachment system.
3. If your machine is variable speed, how do you adjust the speed?
4. What is the primary use of a jig saw in your shop?
5. Who manufactures your machine?

Safety Quiz: True or False

1. Goggles are not necessary when using this machine.
2. Jig saws can be used to do intricate cutting.
3. Fingers must be kept clear of the blade.
4. A teacher's demonstration is not necessary before using this machine.

Teacher Assignment:

I understand the jig saw or scroll saw.   Signed:_____
Date:_____

Grinder

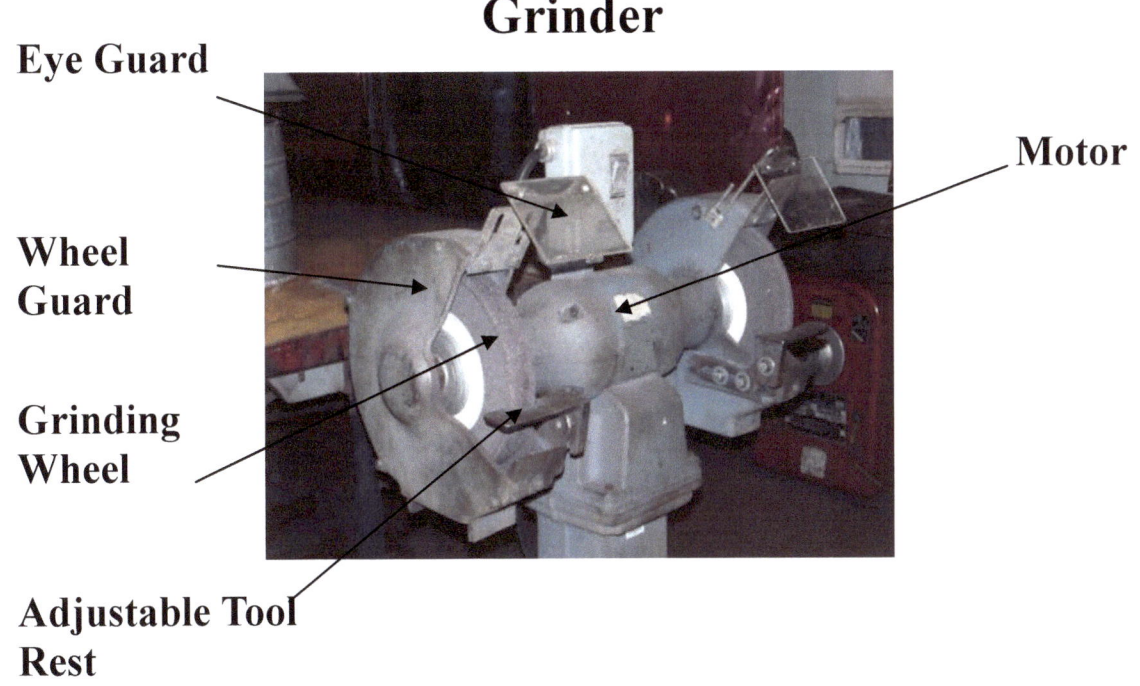

Eye Guard

Wheel Guard

Grinding Wheel

Adjustable Tool Rest

Motor

Safety Rules:

1. Operate only after you have received proper safety instruction from your teacher.
2. No loose clothing.
3. Always wear goggles.
4. Keep fingers away from grinding wheel.
5. The adjustable tool rest should be no more than 1/16" from the grinding wheel.
6. Grind only on the face of the wheel.
7. Make sure all pieces are secured.
8. Be sure floor area around machine is clean and not slippery.
9. If grinding wheels get clogged or out of balance, have your instructor dress the wheels with a grinding wheel dressing tool.
10. Always be sure to wash your hands when finished grinding to prevent transfer of dust and abrasive to your eyes, etc.

Grinder

Review Questions:

1. How far should the adjustable tool rest be from the grinding wheel?
2. Why must a firm grip be maintained on the piece when grinding?
3. How should small pieces be held when grinding?
4. What tool is used for dressing a grinding wheel?
5. When should a grinding wheel be dressed?
6. Who manufactures your machine?

Safety Quiz: True or False

1. Goggles are not necessary when using this machine as it has safety shields.
2. It is a good idea to jam stock into a grinding wheel.
3. The floor area should be clean around the machine.
4. The adjustable tool rest should be no more than 1/16" from the grinding wheel.
5. Grinding on the side of the wheel is safe.

Teacher Assignment:

I understand the grinder   Signed:_____
Date:_____

Power Hack Saw

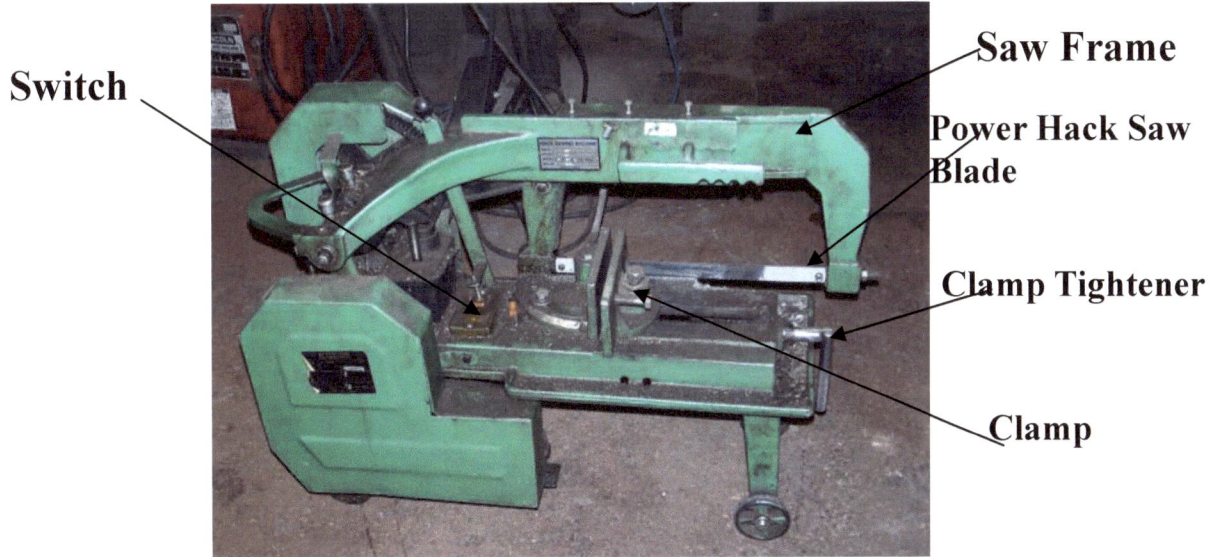

Safety Rules:

1. Operate only after you have received proper safety instruction from your teacher.
2. No loose clothing.
3. Always wear goggles.
4. Be sure stock being cut is clamped securely.
5. Follow your manufacturers recommendations for starting your machine.
6. Never apply pressure to the saw frame to speed cutting.
7. Always keep hands a safe distance from any moving parts.
8. If the saw has a lubrication system, adjust the flow accordingly.
9. Sawing long stock may require additional support.
10. When finished cutting, remove stock from clamp.
11. Never leave the saw running unattended.
12. Clean chips from the machine with a bench brush.
13. Be sure floor area around machine is clean and not slippery.
14. After sawing, remove "burrs" from the cut piece with a file.

Power Hack Saw

Review Questions:

1. Describe how to set up your power hack saw and cut a piece of steel.
2. How does your teacher want you to cut angle iron in your machine?
3. Can you cut tool steel with your machine?
4. Who manufactures your machine?

Safety Quiz: True or False

1. Goggles are not necessary when using this machine.
2. Since this machine is easy to use I can walk away from it
3. The floor around the machine can get slippery from cutting lubricants, so caution must be observed.
4. Sawing long stock may require additional support.

Teacher Assignment:

I understand the power hack saw   Signed:_____
Date:_____

Power Miter Box

Safety Rules:

1. Operate only after you have received proper safety instruction from your teacher.
2. No loose clothing.
3. Always wear goggles.
4. Be sure power is disconnected before making any adjustments.
5. Be sure the guard is functioning properly.
6. Hold stock firmly against the fence and table.
7. Wait until the saw has reached full speed before starting to cut.
8. Be sure floor area around machine is clean and not slippery.

Power Miter Box

Review Questions:

1. Why is it a good idea to check the guard to be sure it is working properly?
2. Why must the blade reach full speed before cutting?
3. Describe how to crosscut a board.
4. What is the maximum size of a board that can be crosscut on your machine?
5. Who manufactures your machine?

Safety Quiz: True or False

1. Goggles are not necessary when using this machine.
2. This machine is used for ripping lumber.
3. This machine can be used to cut angles on stock such as a picture frame.
4. A sharp blade is not important.

Teacher Assignment:

I understand the power miter box. Signed:_____
Date:_____

Portable Hand Drill

Safety Rules:

1. Operate only after you have received proper safety instruction from your teacher.
2. No loose clothing.
3. Always wear goggles.
4. Remove the chuck key prior to operation.
5. On drills with a friction chuck make sure twist drill is secure before starting.
6. Make sure all pieces being drilled are secured.
7. Be sure floor area around machine is clean and not slippery.
8. Keep a firm grip on the drill while drilling.
9. Use sharp twist drills.

Hand Drills

Review Questions:

1. Why must the twist drill be tight in the chuck?
2. Why must a firm grip be maintained when using this tool?
3. Why must pieces be clamped down while drilling?
4. Describe how to chuck a twist drill in your machine.
5. Who manufactures your machine?

Safety Quiz: True or False

1. Goggles are not necessary when using this machine.
2. A loose T shirt is OK when operating this machine.
3. Small pieces being drilled can be hand held.
4. Sharp twist drills are important.

Teacher Assignment:

I understand the hand drill   Signed:_____
Date:_____

Router

Depth Adjustment

Collet Type Chuck

Switch

Handle

Base

Safety Rules:

1. Operate only after you have received proper safety instruction from your teacher.
2. No loose clothing.
3. Always wear goggles.
4. Be sure to unplug router before making any adjustments.
5. Be sure router bit is secure in collet chuck.
6. Work piece should be secure.
7. Routers require 2 hand operation.
8. Always make a test cut on scrap wood to see if bit depth is OK and correct bit is being used.
9. Make sure switch is in the "off" position before plugging in.
10. When finished routing, wait until the router stops turning before setting it down.
11. Use sharp router bits.
12. Be sure floor area around machine is clean and not slippery.
13. See your instructor for the maximum permissible depth cut allowed on your machine and proper speed rate for your material being routed..

Router

Review Questions:

1. What is the purpose of a router?
2. Why is it a good idea to make a test cut first?
3. Why should you check the power switch before plugging this machine in?
4. Who manufactures your machine?

Safety Quiz: True or False

1. Router bits must be secure in the collet chuck.
2. Routers can be used with one hand.
3. Routers can be used to put special edges on boards.
4. Routers spin at very high RPMs.

Teacher Assignment:

I understand the router.   Signed:_____
Date:_____

Portable Belt Sander

Safety Rules:

1. Operate only after you have received proper safety instruction from your teacher.
2. No loose clothing.
3. Always wear goggles.
4. Unplug machine when changing belts.
5. Be sure power switch is in the off position before plugging in.
6. Use caution when adjusting belt tracking.
7. Use both hands when operating the belt sander.
8. Have the sander running before applying to the wood.
9. Be sure belt sander is stopped before setting it down.
10. Make sure all pieces are secured.
11. Be sure floor area around machine is clean and not slippery.
12. Change belts when they are worn out.

Belt Sander

Review Questions:

1. Why is it a good idea to be sure the trigger switch is in the off position before plugging this machine in?
2. How do you adjust the tracking on your machine?
3. Why must stock be secured when sanding?
4. Explain differences in grit sizing on the abrasive belt. For example, is a #60 grit belt, coarser or finer than a #120?
5. Who manufactures your machine?

Safety Quiz: True or False

1. Goggles are not necessary when using this machine.
2. A belt sander is a one hand machine.
3. Worn out belts must be replaced.
4. Change belts after unplugging the machine.

Teacher Assignment:

I understand the belt sander. Signed:_____
Date:_____

Portable Electric Circular Saw

Trigger Switch
Handle
Retractable Guard
Tilting Base
Blade

Safety Rules:

1. Operate only after you have received proper safety instruction from your teacher.
2. No loose clothing.
3. Always wear goggles.
4. Be sure the retractable guard is working properly.
5. The saw should reach full speed before cutting.
6. Be sure of cutting saw path.
7. When cutting, do not twist the saw.
8. Any adjustments should be made with the saw unplugged.
9. If while cutting, the blade binds, stop immediately.
10. Use a firm grip while operating the saw.
11. Do not set the saw down until the blade stops.
12. Make sure all pieces are secured.
13. Be sure floor area around machine is clean and not slippery.

Portable Electric Saw

Review Questions:

1. Why must the guard work?
2. Why do you have to check the underside area below the stock being cut ?
3. Why is a clean floor area important?
4. Why must a firm grip be maintained when using this saw?
5. Who manufactures your machine?

Safety Quiz: True or False

1. Goggles are not necessary when using this machine.
2. The floor area should be clean around the machine.
3. Portable electric saws can be dangerous if used improperly.
4. Setting the saw down while it is running is OK.

Teacher Assignment:

I understand the portable electric saw.   Signed:_____
Date:_____

Planer-Surfacer

Safety Rules:

1. Operate only after you have received proper safety instruction from your teacher.
2. No loose clothing.
3. Always wear goggles.
4. Stock being surfaced should be free of loose knots, nails, paint or any other foreign objects.
5. Select the proper depth of cut according to the manufacturer's recommendation.
6. Do not look into the throat area while the planer is running.
7. Do not force stock through the planer.
8. Keep hands away from the throat area.
9. Remove all chips with a bench brush
10. Be sure floor area around machine is clean and not slippery.
11. See your instructor for the maximum permissible cut allowed on your machine.

Planer-Surfacer

Review Questions:

1. Why is it important to not have loose knots, excessive glue on glue joints, nails, paint, etc. on wood being surfaced?
2. What is the maximum permissible cut on your machine?
3. What is the shortest board allowed on your machine?
4. How is the size of your planer indicated?
5. What should you do should a board become stuck in your planer?
6. What is the thinnest your planer can surface a board to?
7. Who manufactures your machine?

Safety Quiz: True or False

1. It is a good idea to look into the throat of a running planer.
2. Goggles are mandatory while running a planer.
3. Clean chips from the machine with your hands.
4. The main purpose of a planer-surfacer is to plane edges of lumber.

Teacher Assignment:

I understand the planer-surfacer.
Signed:_____ Date:_____

Electric Arc Welder

Switch

Electrode Holder

Ampere Selector

Ground Clamp

Safety Rules:

1. Operate only after you have received proper safety instruction from your teacher.
2. No loose or flammable clothing.
3. Always wear goggles when chipping slag
4. Always use a face shield with a minimum #10 lens when arc welding to protect from ultraviolet rays
5. Wear clothing that will resist sparks.
6. Be sure your clothing will protect all body parts from ultraviolet rays.
7. Wear welding gloves.
8. Have good ventilation in the welding area.
9. Warn others in your welding area before welding.
10. Be sure your welding area is free of combustible materials.
11. Be sure floor area around machine is clean and not slippery.

Electric Arc Welder

Review Questions:

1. Why is it important to not have combustible material around an arc welder?
2. Why is adequate ventilation important?
3. What are the rays given off while arc welding and why must one protect themselves from them?
4. What is the minimum number lens shade for arc welding?
5. What is slag?
6. What is welders flash?
7. Who manufactures your machine(s)?

Safety Quiz: True or False

1. A number 5 lens is sufficient for arc welding.
2. Sunglasses provide adequate protection while arc welding.
3. Gloves are required when arc welding
4. All exposed skin must be covered when arc welding.

Teacher Assignment:

I understand the arc welder.   Signed:_____
Date:_____

Oxyacetylene Welding Equipment

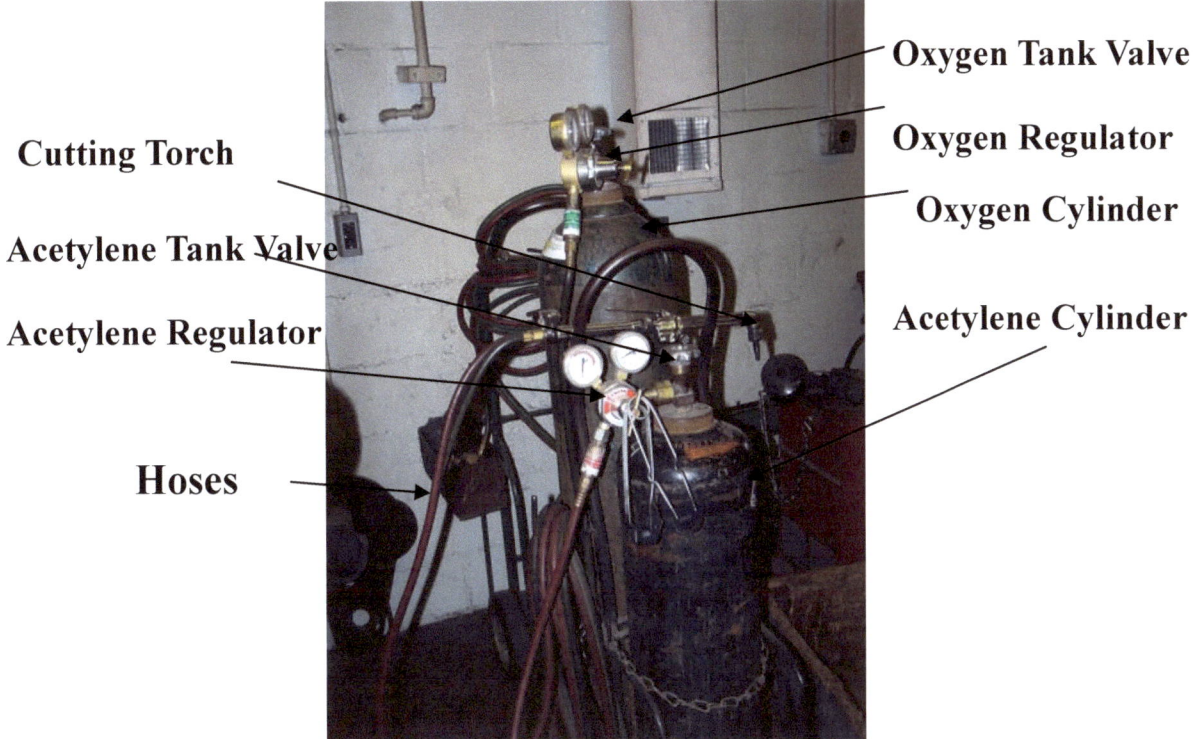

Safety Rules:

1. Operate only after you have received proper safety instruction from your teacher.
2. No flammable clothing.
3. Always wear cutting goggles with at least a # 5 lens.
4. Open acetylene tank valve no more than 1/4 of a turn.
5. Open oxygen tank valve all the way.
6. When shutting down, bleed all gases out of hoses and from regulators.
7. Wear welding gloves.
8. Never lay down a lit torch.
9. Keep all petroleum based products away from oxygen equipment.
10. Acetylene tanks should be in an upright position.
11. Acetylene working pressure should never exceed 15 psi.
12. Have good ventilation in the welding area.
13. Be sure your welding area is free of combustible materials.
14. Check equipment often for leaks.
15. Follow all manufacturers usage instructions.

Oxyacetylene Welding Equipment

Review Questions:

1. What number lens is used for oxyacetylene welding?
2. Why is it important not to exceed 15 psi acetylene line pressure?
3. Why is oil dangerous around oxygen?
4. Why is it important to check for leaks?
5. Why is it important to follow your equipment manufacturers lighting and shutdown procedures?
6. Who manufactures your equipment?

Safety Quiz: True or False

1. It is OK to cut with 50 psi of acetylene.
2. Never wear greasy, oily clothes while oxyacetylene welding.
3. This kind of welding requires no gloves.
4. Oxyacetylene cylinders should be chained in an upright position.

Teacher Assignment:

I understand oxyacetylene equipment.   Signed:_____
Date:_____

Disk Sander

Safety Rules:

1. Operate only after you have received proper safety instruction from your teacher.
2. No loose clothing.
3. Always wear goggles.
4. Sand only on the downward side.
5. Use near an exhaust system.
6. Remove dust with a bench brush
7. Be sure floor area around machine is clean and not slippery.

Disk Sander

Review Questions:

1. Which direction does your sander rotate?
2. Which side of the machine is the downward side?
3. Why is dust collection equipment important?
4. Who manufactures your machine?

Safety Quiz: True or False

1. Change abrasive disks when they are worn out.
2. Sand only on the downward side (rotation) on your disk sander.
3. Do not wear goggles.
4. Use the disk sander as a saw and remove large amounts of material with it.

Teacher Assignment:

I understand the disk sander   Signed:_____
Date:_____

Vertical Milling Machine

Safety Rules:

1. Operate only after you have received proper safety instruction from your teacher.
2. No loose clothing.
3. Always wear goggles.
4. Be sure the cutter is sharp and tightly held in the chuck and the material being milled is secured.
5. Keep hands away from the cutter.
6. Remove chips with a brush after the machine is stopped.
7. Be sure floor area around machine is clean and not slippery.
8. Check all adjustments such as spindle rotation, speed, depth of cut before starting the cut.
9. Once you have made a cut, check proper clearance before the next cut

Vertical Milling Machine

Review Questions:

1. Why must the cutter be tight?
2. Why is eye protection important when using this machine?
3. What is the maximum permissible cut on your machine for steel? For aluminum?
4. Who manufactures your machine?

Safety Quiz: True or False

1. Make adjustments only when the machine is off.
2. The work piece must be secured.
3. A dull cutter is OK.
4. If the machine is running slow, goggles are not necessary.

Teacher Assignment:

I understand the vertical milling machine.   Signed:_____
Date:_____

Horizontal Band Saw

Safety Rules:

1. Operate only after you have received proper safety instruction from your teacher.
2. No loose clothing.
3. Always wear goggles.
4. All adjustments are made with the power off.
5. Be sure blade guides are adjusted properly.
6. Be sure work is clamped tightly.
7. Adjust feed rate properly.
8. Remove any chips with a bench brush.
9. Remain at the machine while it is running.

Horizontal Band Saw

Review Questions:

1. Why is it important to secure the work?
2. What should be used to remove chips?
3. How should the blade guides be adjusted on your machine?
4. What does a knocking sound from the blade indicate?
5. Who manufactures your machine?

Safety Quiz: True or False

1. It is safe to put hand pressure on the saw frame when cutting.
2. Work can be put loosely in the machine and securing it is not important..
3. Goggles are necessary when using this machine.
4. It is a good idea to talk to others when using this machine.
5. If a blade breaks, alert your teacher immediately.

Teacher Assignment:

I understand the band saw.   Signed:_____
Date:_____

Teacher Copy   Common Hand Tools

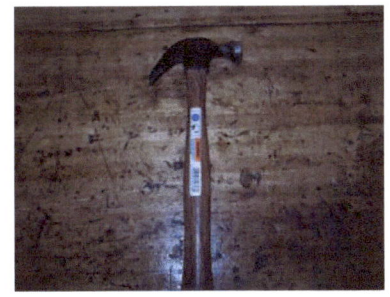

Claw Hammer

Ball Peen Hammer

Combination Square

Flat File

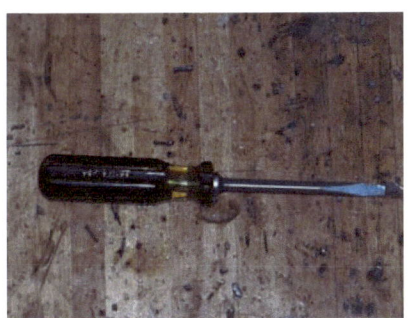

Flat Screwdriver

Framing Square

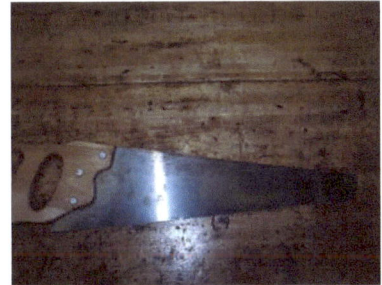

Hand Saw

Level

Common Hand Tools

1.

2.

3.

4.

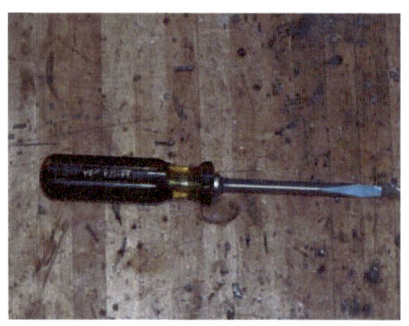

5.

6.

7.

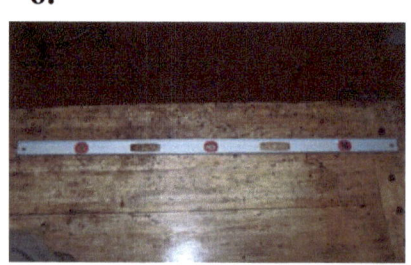

8.

Teacher Copy   Common Hand Tools

Needle Nose Pliers

Phillips Screwdriver

Rubber Mallet

Tape Measure

Twist Drill

Wood Chisel

Wrecking Bar

Common Hand Tools

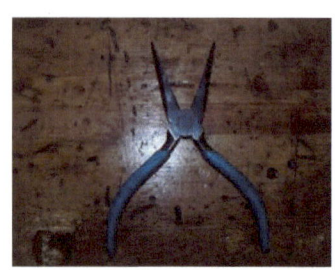

9.

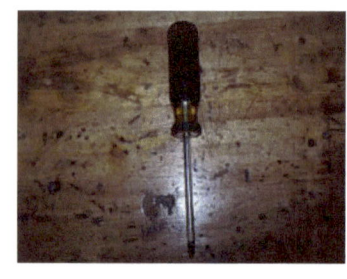

10.

11.

12.

13.

14.

15.

Power Tool Safety Exam   Name_____
                              Date_____
                              Student initials_____

Answer: True or False

1. Running a lathe requires your undivided attention. _____
2. Holding pieces in your hand while drilling is OK. _____
3. It is safe to crosscut with the miter gauge. _____
4. Jointers can be used to joint edges of wood. _____
5. A jointer's guard must be working. _____
6. Crosscutting can be done on a radial arm saw. _____
7. The top guide should be 1" from the piece being cut on a band saw. _____
8. Jig saws can be used to do intricate cutting. _____
9. The adjustable tool rest should be no more than 1/16" from the grinding wheel. _____
10. The fence is not necessary when ripping stock on a table saw. _____
11. The floor around the power hack saw can get slippery from cutting lubricants so caution must be observed. _____
12. Goggles are not necessary when using power tools. _____
13. The power miter box is used for ripping lumber. _____
14. Remove chuck keys from the drill press before starting. _____
15. Routers spin at high RPMs. _____
16. It is mandatory the guard works on a portable circular saw. _____
17. It is OK to wear a #1 lens when oxyacetylene welding. _____
18. The minimum lens for arc welding is a #10. _____
19. Horseplay is never permitted in a shop. _____
20. A clean and organized shop represents a level of professionalism. _____
21. Clean floors are important around power equipment. _____
22. Never use any machine until you have watched a demonstration by your teacher and have their permission to use the machine. _____
23. Shop safety is of utmost importance. _____
24. Safety guards usually get in the way and therefore can be removed. _____
25. Special setups on any machine require the instructors approval and inspection. _____

Teacher Safety Exam Key

1. T
2. F
3. T
4. T
5. T
6. T
7. F
8. T
9. T
10. F
11. T
12. F
13. F
14. T
15. T
16. T
17. F
18. T
19. T
20. T
21. T
22. T
23. T
24. F
25. T

References:

Cooper, E. (1987) *Agriculture Mechanics: Fundamentals and Applications*. Albany, NY: Delmar Publishers Inc.

Feir, J. L. (1963) *Advanced Woodwork and Furniture Making*. Peoria, Illinois: Chas. A. Bennet Co., Inc PUBLISHERS

Mincemoyer, D. L., Williams, W., & Curtis, S. (1974) *Safe Power Shop Equipment Operation*. University Park, PA: The Pennsylvania State University

Pynnonen, A., *Power Tool Manual*. Sheldon, WI: National Vo-Ag Book Company

http://www.agednet.com

http://www.powertoolinstitute.com/

www.ingramcontent.com/pod-product-compliance
Lightning Source LLC
Chambersburg PA
CBHW041523220426
43669CB00002B/33